History of quantum physics: The origins of a new revolution

Maxime STONE

Collection:
The keys to understanding.

Copyright © 2023 Maxime STONE
All rights reserved.
ISBN: 9798879199413

Preface

As I embark on this fascinating journey through the mysteries of quantum physics, I would like to begin by expressing my deep gratitude to two key bodies who have made this exploration possible: the French Atomic Energy Commission (CEA) and the physicist, philosopher of science and radio producer Étienne Klein.

The CEA, with its unwavering commitment to scientific research and innovation, has played a crucial role in democratising science and promoting public understanding of the most complex and fascinating areas of physics. His unwavering support for fundamental and applied research has not only enriched the scientific community, but has also enabled science enthusiasts like me to delve into the depths of quantum physics and explore its mysteries.

As for Étienne Klein, his contribution to this field cannot be overstated. A renowned physicist, enlightened philosopher of science and outstanding science communicator, Mr Klein has a unique ability to illuminate the most obscure concepts of quantum physics and make them accessible to a wider audience. His passion for popularising science, combined with a deep philosophical understanding of the implications of quantum physics, has been a constant source of inspiration throughout the writing of this book.

This book is an attempt to navigate through the complex and often bewildering labyrinth of quantum physics, a field that defies our traditional conceptions of the world and continues to astonish even the most erudite of scientists. By exploring its principles, applications and implications, we discover not only the foundations of the universe, but also confront profound philosophical questions about the nature of reality, the structure of space and time, and our place in this immense cosmos.

With thanks to the CEA for its pioneering role in scientific research and to Étienne Klein for his ability to make science both profound and accessible, I invite you to immerse yourself in the pages that follow. Whether you are a keen student, a seasoned researcher or simply curious about the world of science, I hope that this book will offer you not only knowledge, but also a sense of wonder and a renewed appreciation for the enigmatic and beautiful world of quantum physics.

So, let's take a fascinating journey into the heart of one of the most mysterious and inspiring areas of modern science.

With all my consideration,

Maxime.

Table of contents

Preface iii
Chapter 1: Introduction to Quantum Physics 7
Chapter 2: History of Quantum Physics 11
Chapter 3: Fundamental Concepts of Quantum Physics 14
Chapter 4: The Emergence of Quantum Mechanics: 17
Chapter 5: The Key Figures of Quantum Physics 21
Chapter 6: The Theory of Relativity and Quantum Physics 26
Chapter 7: Applications and Implications of Quantum Physics. 29
    1. Technology and Electronics 29
    2. Quantum Computing 30
    3. Medicine and Imaging 30
    4. Materials Science 30
    5. Energy and Environment 31
    6. Philosophical and Cultural Impact 31
    7. Emerging Technologies 31
Chapter 8: Challenges and controversies 33
    1. Indeterminism and Predictability 33
    2. The Role of the Observer 33
    3. Quantum entanglement and non-locality 34
    4. Interpretations of Quantum Mechanics 34
    5. Limits of Current Understanding 35
    6. Impact on other areas 35
Chapter 9: Quantum Physics and Philosophy 37
    1. Reality and Objectivity 37
    2. Causality and Determinism 37

3. Knowledge and understanding 38
4. Metaphysical implications 38
5. Global Philosophical Impact 39
6. Ethics and Responsibility 39

Chapter 10: The Future of Quantum Physics 41
1. Unification of Fundamental Theories 41
2. Quantum Technologies 41
3. Advances in Particle Physics and Cosmology 42
4. Interdisciplinary Applications 42
5. Philosophical and ethical challenges 43
6. Education and Society 43

Epilogue: On the threshold of a new horizon 45
References and additional resources 48
1. Books and articles 48
2. Online resources 49
3. Documentaries and Educational Videos 49

Chapter 1: Introduction to Quantum Physics

Quantum physics is a branch of science that has revolutionised our understanding of the universe at the microscopic level. This field, which emerged at the beginning of the 20th century, has not only challenged the laws of classical physics, but has also opened up new horizons in science and technology.

Unlike classical physics, which describes the world on a visible, macroscopic scale, quantum physics focuses on phenomena that take place on the scale of atoms and subatomic particles. This microscopic scale reveals a universe where the laws of classical physics no longer apply in the same way, introducing concepts such as wave-particle duality, superposition and quantum entanglement.

- Wave-particle duality, eloquently illustrated by the double-slit experiment, suggests a reality in which the categorical distinctions between waves and particles become blurred. This concept has profound implications for our understanding of light and matter, challenging the very foundations of classical physics. This duality is at the heart of many modern technologies, including semiconductors and lasers, and continues to inspire innovative research in fields such as photonics and the mechanics of materials.

- Superposition, strikingly evoked by Schrödinger's cat, reveals a world where reality is not fixed but probabilistic right up to the act of measurement. This intrinsic characteristic of quantum mechanics opens the way to revolutionary applications such as quantum computing, which promises to solve previously insurmountable computational problems. By exploiting the ability of qubits to exist in multiple states simultaneously, quantum computing could transform fields ranging from cryptography to molecular modelling.

- Quantum entanglement defies any classical understanding of space and time. This phenomenon, in which particles remain connected regardless of the distance separating them, suggests a level of correlation that transcends the limits of classical communication. Entanglement lies at the heart of quantum cryptography, which promises secure communications that cannot be intercepted or decrypted by conventional means. It also opens up exciting prospects for quantum networks, which could revolutionise the way we share and process information.

Quantum physics, with its revolutionary approach and astonishing discoveries, has not only shaped the world of science and technology, but has also sparked a philosophical awakening to fundamental questions about our existence. Its implications reach far beyond laboratories and equations to touch the very essence of our understanding of reality.

Technologies derived from quantum physics, such as transistors and lasers, are the cornerstone of the modern digital age. They have revolutionised communication and information processing and paved the way for the era of advanced computing. Nuclear

History of quantum physics: The origins of a new revolution

Magnetic Resonance (NMR), another major application, has transformed the medical field by providing a non-invasive means of exploring the inside of the human body, greatly improving the diagnosis and treatment of disease.

Although still in its infancy, quantum computing promises to revolutionise our ability to process and analyse massive amounts of data. With its potential to solve problems otherwise unsolvable by conventional computers, it could open up new perspectives in medical research, cryptography, materials design and much more.

Quantum physics has not only pushed back the frontiers of technology; it has also prompted philosophers and scientists to rethink fundamental concepts. By shattering the notions of causality and objective reality, it has opened up a new field of philosophical questioning. The indeterminacy inherent in quantum mechanics, as well as paradoxes such as Schrödinger's cat, have stimulated discussions about the nature of reality, the role of the observer and the very concept of knowledge.

By redefining our understanding of the universe at the microscopic level, quantum physics has not only broadened our vision of the physical world but has also enriched our understanding of our place in the universe. It raises questions about the nature of existence and the structure of the universe, prompting a deeper exploration of these mysteries.

By laying the foundations of quantum physics, this first chapter opens the door to a fascinating world of possibilities and discoveries. It represents a starting point for an intellectual adventure that continues to transform our world, revealing the hidden threads that weave the fabric of our reality. Quantum physics remains a fertile field for scientific discovery and an inexhaustible source of inspiration for philosophical questioning.

History of quantum physics: The origins of a new revolution

Chapter 2: History of Quantum Physics

Quantum physics, with its rich historical legacy and revolutionary contributions, is a fascinating story of discovery and innovation. By tracing its origins and development, we discover a series of intellectual breakthroughs that have profoundly changed our understanding of the world.

The history of quantum physics began at the turn of the twentieth century. In 1900, Max Planck, faced with the problem of blackbody radiation, proposed a radical solution: energy is emitted in discrete quantities, or quanta. This fundamental idea marked the birth of quantum theory.

Albert Einstein continued this revolution in 1905 with his theory of the quanta of light, explaining the photoelectric effect. This work established that light behaves both as a wave and as a particle, a concept that defied classical understanding.

In the 1920s, quantum mechanics rapidly developed into a coherent theory thanks to the work of several scientists. In 1925, Werner Heisenberg formulated matrix quantum mechanics, providing a new perspective on the quantum states of particles.

Almost simultaneously, Erwin Schrödinger developed the wave equation that bears his name, providing a wave description of quantum mechanics. Schrödinger's equation has become one of the pillars of quantum physics, describing how the quantum state of a system evolves over time.

History of quantum physics: The origins of a new revolution

Niels Bohr, another major player in this period, introduced the Bohr model of the atom in 1913, which explained the emission spectra of atoms. Bohr also played a central role in developing the Copenhagen interpretation of quantum mechanics, which remains one of the most widely accepted interpretations today.

In 1928, Paul Dirac proposed the Dirac equation, which reconciled quantum mechanics with special relativity and predicted the existence of antimatter. This discovery paved the way for research into particle physics.

The formative period of quantum physics, punctuated by intellectual debates and revolutionary discoveries, represents a fascinating chapter in the history of science. The exchanges between emblematic figures not only shaped the development of quantum physics, but also highlighted the profound philosophical implications of this theory.

The debate between Einstein and Bohr has become emblematic of the tension between two visions of the world: one classical and deterministic, the other quantum and probabilistic. Einstein, with his conviction that "God doesn't play dice", was looking for a more determined and predictable reality. Bohr, on the other hand, embraced the intrinsically probabilistic and indeterminate nature of quantum mechanics. This confrontation illuminated the complexity of quantum physics and revealed the extent to which our intuitions can be challenged by the fundamental laws of nature.

The impact of quantum physics extends far beyond theoretical debates. It has profoundly transformed our understanding of fundamental phenomena such as atomic structure, the nature of light and energy. Its practical applications have revolutionised fields as varied as chemistry, medicine, materials and, above all, information technology. The miniaturisation of electronic components, advances in medical imaging and the development of new methods of cryptography are just a few examples of its

History of quantum physics: The origins of a new revolution

considerable influence.

The history of quantum physics has been marked by a series of bold innovations and surprising discoveries. Each breakthrough has opened up new questions, stimulating ongoing exploration into the depths of the atom and beyond. These discoveries are not just intellectual triumphs; they represent an ongoing expansion of the frontiers of human knowledge.

By unravelling the mysteries of the atom and light, quantum physicists have also posed fundamental questions about the nature of reality. These questions continue to provoke intense debate and stimulate research in emerging areas of physics. Current theories, while powerful, are not yet complete, and the quest for a deeper understanding of the universe continues.

Quantum physics is a never-ending intellectual adventure, a field of constant discovery that continues to push back the boundaries of our understanding and challenge our imagination. The legacy of quantum physics is an eloquent testament to human curiosity and our relentless quest to understand the very foundations of our reality.

Chapter 3: Fundamental Concepts of Quantum Physics

We are going to deepen our understanding of the fundamental principles that form the bedrock of quantum physics. These principles, which are radically different from those of classical physics, have opened up a new field of scientific understanding.

At the heart of quantum physics is the idea of <u>indeterminism</u>, a radical departure from classical determinism. In the quantum world, the behaviour of particles is intrinsically unpredictable and probabilistic. This means that, unlike classical mechanics, where the future of a system can be accurately predicted if its current state is known, quantum physics allows only probabilistic predictions.

The <u>principle of superposition is one of</u> the most enigmatic aspects of quantum physics. According to this principle, a quantum particle, such as an electron, can exist in several states or configurations simultaneously. It is only when it is measured or observed that the particle 'chooses' a specific state. This probabilistic nature manifests itself in phenomena such as Schrödinger's cat model, where a system can be in several states simultaneously.

<u>Measurement in quantum physics</u> is a unique process. When a measurement is made on a quantum system, it forces the system to 'decide' on a particular state from among all the superimposed possibilities. This process, often called 'wave packet reduction',

remains one of the most debated and mysterious aspects of quantum theory, raising questions about the role of the observer and the nature of reality.

Wave-particle duality is another key concept. It states that quantum particles, such as electrons and photons, can behave both as particles and as waves. This duality is illustrated by the double slit experiment, where individual particles passing through two adjacent slits create a wave interference pattern on a screen, a typically wave-like behaviour.

Quantum entanglement is perhaps the strangest and most fascinating phenomenon in quantum physics. In a state of entanglement, two or more particles become so tightly bound that the state of one instantaneously affects the state of the other, no matter how far apart they are. This defies our intuitive understanding of locality and has profound implications for information transmission and quantum information theory.

Heisenberg's uncertainty principle states that it is impossible to measure the position and velocity (or momentum) of a particle accurately at the same time. The more accurately one of these properties is known, the less accurately the other is known. This principle underlines the fundamentally probabilistic nature of quantum mechanics and has important implications for our understanding of reality on the quantum scale.

When a measurement is made on a quantum system, its wave function 'collapses' to a specific state. This collapse of the wave function is central to many interpretations of quantum mechanics, with profound philosophical implications for the nature of reality and the role of the observer.

The fundamental concepts of quantum physics define a universe that is both fascinating and disconcerting, where reality does not behave as we would expect according to classical laws. We have just explored the fundamental ideas at the heart of this revolutionary theory, revealing how they challenge our established conceptions of reality and pave the way for new advances in science and technology.

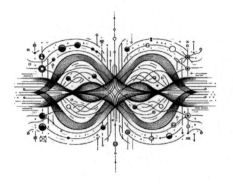

Chapter 4: The Emergence of Quantum Mechanics:

Quantum mechanics, one of the greatest achievements of modern physics, was born out of a series of daring experiments and theories in the early twentieth century. Let's explore the key advances and experiments that paved the way for this revolutionary formulation.

At the beginning of the twentieth century, science was faced with a series of mysteries that challenged the established understanding of physics. Despite the remarkable progress of classical physics, epitomised by the work of Newton and Maxwell, certain phenomena remained unexplained, posing serious challenges to the very foundations of physics.

One of the most enigmatic problems was blackbody radiation. According to classical physics, a black body - an ideal object that absorbs all incident light energy - should emit radiation whose intensity increases indefinitely with frequency, a prediction known as the "ultraviolet catastrophe". However, experiments showed a very different spectrum of radiation, suggesting that the classical understanding of radiant energy was fundamentally flawed. This mystery raised fundamental questions about the nature of light and energy.

Another puzzling phenomenon was the photoelectric effect, observed when light strikes certain materials and releases electrons. According to the theories of light at the time, the intensity of the light should affect the energy of the ejected electrons. However, experiments showed that it was the frequency of the light, not its intensity, that determined this energy. This behaviour contradicted the classical understanding of light as a continuous wave, suggesting that light also had particle properties.

The spectral lines of the atoms were also a source of puzzlement. The emission and absorption spectra of atoms showed lines at specific wavelengths, indicating discrete energy transitions. However, classical theory could not explain why atoms emitted or absorbed energy at these specific wavelengths, rather than at a continuous spectrum.

These mysteries highlighted the limits of classical physics and prompted scientists to look for new explanations. This period of intense questioning and research laid the foundations for the quantum revolution. Scientists were on the verge of discovering that the microscopic universe obeyed very different rules to those of the macroscopic world, opening up a new era in our understanding of the nature of reality.

The key experiments in the development of quantum mechanics have not only shed light on previously unknown aspects of nature, but have also challenged the foundations of our understanding of reality. These experiments have played an essential role in the consolidation and acceptance of quantum mechanics.

The double slit experiment, first carried out with light and later with electrons, has become emblematic of the strange nature of quantum physics. When individual particles, such as electrons, are shot through two parallel slits, they produce an interference pattern on a screen behind the slits, similar to that created by waves. When one of the slits is closed, the interference pattern disappears, suggesting that each electron is interfering with itself as if it were passing through both slits simultaneously. This experiment clearly demonstrates wave-particle duality, where particles can exhibit both particle and wave properties, depending on the experiment.

Quantum entanglement experiments, in particular those testing the EPR paradox and Bell's inequalities, have provided irrefutable proof of one of the most non-intuitive aspects of quantum mechanics: entanglement. In these experiments, two particles are prepared in such a way that their quantum states are linked, so that measuring the state of one particle immediately reveals information about the state of the other, regardless of the distance separating them. These results, which appear to violate the limits of the speed of light for causal influence, suggest that entangled particles share a kind of connection that transcends traditional space and time.

These experiments were crucial not only for the validation of quantum theory, but also for stimulating profound philosophical and scientific debates about the nature of reality. The double slit experiment, in particular, challenged the idea of an objective, well-defined reality independent of observation. It raised questions about the role of the observer in determining the results of an experiment.

Similarly, quantum entanglement experiments have challenged notions of locality and separation, suggesting that the universe is much more interconnected than previously thought. They have paved the way for new theories and applications, notably in quantum computing and quantum cryptography, where entanglement is used for information-processing tasks that are impossible by conventional standards.

These key experiments, by challenging our perceptions and broadening our understanding of the nature of matter and light, have been milestones in the evolution of quantum physics. They have not only confirmed counter-intuitive aspects of the theory, but have also led to new questions and discoveries, continuing to

History of quantum physics: The origins of a new revolution

shape our exploration of quantum reality.

Chapter 5: The Key Figures of Quantum Physics

Quantum physics, as a revolutionary field of science, owes its development and maturation to a group of brilliant scientists whose individual contributions have profoundly shaped modern understanding of the subatomic world. Let's explore the lives and achievements of some of the key figures in quantum physics: Albert Einstein, Niels Bohr, Erwin Schrödinger, and Werner Heisenberg.

Albert Einstein: Pioneer of Relativity and Critic of Quantum Mechanics

Albert Einstein (1879-1955) is often associated with the theory of relativity, but his contributions to quantum physics are also major. In 1905, he explained the photoelectric effect, asserting that light is made up of quanta, later called photons. This discovery, for which he was awarded the Nobel Prize, was fundamental in establishing quantum theory.

Albert Einstein had a passion for music, particularly the violin. He often said that if he hadn't been a scientist, he would have been a musician. Music was a source of joy and inspiration for him. Once, when asked to explain his theory of relativity in an understandable way, he replied: "When I sit at my desk and work on my theories for a few hours, it seems like a few minutes. But when I sit down and play the violin for a few hours, it seems like a few minutes. That's relativity."

History of quantum physics: The origins of a new revolution

Niels Bohr: The Atomic Model and Quantum Philosophy

Niels Bohr (1885-1962), a Danish physicist, is famous for his 1913 model of the atom, which introduced quantized orbits for electrons. This model helped explain why atoms emit light at specific wavelengths. Bohr also played a key role in developing the Copenhagen interpretation of quantum mechanics, which holds that particles do not have defined properties independent of measurement. His vision has profoundly influenced the philosophy and understanding of quantum physics.

Niels Bohr was known for his rigorous scientific mind, but a famous story reveals a lighter side to his personality. A colleague was surprised to see a horseshoe hanging above Bohr's front door, an object often associated with superstition. When asked if he really believed that the horseshoe would bring him luck, Bohr reportedly quipped, "Of course not, but I've been told it works whether you believe it or not."

Erwin Schrödinger: The Wave Equation and Schrödinger's Cat

Erwin Schrödinger (1887-1961), an Austrian physicist, is best known for his development in 1926 of the wave equation that bears his name. Schrödinger's equation is a fundamental pillar of quantum mechanics, describing how the wave function of a system evolves over time. He is also famous for his "Schrödinger's cat" thought experiment, designed to illustrate the apparent absurdity of quantum superposition on a macroscopic scale.

The most famous anecdote about Erwin Schrödinger is linked to his "Schrödinger's cat" thought experiment. It was an ironic response to the bizarre interpretations of quantum mechanics. Schrödinger imagined a cat locked in a box with a mechanism that could randomly release poison. According to quantum mechanics, the cat would be both dead and alive until someone opened the box. This thought experiment was intended to illustrate the absurdity of applying quantum laws to the macroscopic world and was not to be taken literally.

Werner Heisenberg: The Uncertainty Principle and Matrix Mechanics

Werner Heisenberg (1901-1976), a German physicist, introduced the uncertainty principle in 1927, asserting that it is impossible to measure both the position and velocity of a particle accurately at the same time. This principle is a cornerstone of quantum mechanics, underlining the intrinsically probabilistic nature of measurement on the quantum scale. In 1925, Heisenberg also developed matrix quantum mechanics, a formalised approach to dealing with quantum systems.

Werner Heisenberg met Einstein in 1926. At this meeting, Heisenberg presented Einstein with his uncertainty principle. Einstein, sceptical, replied with his famous counter-argument: "God doesn't play dice". Heisenberg, then young and impressionable, was somewhat taken aback by this encounter with one of the greatest scientists of the time. This interaction highlighted the differences in thinking between the pioneers of quantum physics and those of classical physics, as well as the mutual respect that existed despite their disagreements.

These scientists were not only accomplished physicists; they were also philosophers in their own right, seeking to understand the deeper implications of their discoveries. Their work and debates shaped quantum physics and had a lasting impact on science, philosophy and technology.

1. **Debates and Dialogues**: The interactions between these great minds, particularly the debates between Einstein and Bohr, were crucial in refining and testing the ideas of quantum mechanics. These dialogues helped to clarify and challenge the concepts of the theory, leading to a deeper understanding of its foundations.

2. **Influence on Science and Technology**: Their discoveries have not only broadened our understanding of the universe at a fundamental level, but have also led to practical advances. Technologies such as lasers, transistors and nuclear magnetic resonance, which have transformed modern society, have their roots in quantum mechanics.

3. **Philosophical implications**: The work of these physicists has also had a significant impact on the philosophy of science. They challenged notions of reality, causality and objectivity, and introduced new ways of thinking about the relationship between the observer and the observed system.

The key figures in quantum physics have not only made major scientific contributions, but have also profoundly influenced the way we perceive and interact with the world at its most basic level. Their work has opened doors to new theories and technologies, while challenging our preconceptions about reality and observation. Their collaborative efforts and intellectual debates were essential in refining and deepening our understanding of quantum mechanics, a field that continues to fascinate and challenge scientists and philosophers today. This enduring legacy is a testament to their genius and vision, laying a solid foundation on which modern physics continues to build and develop.

History of quantum physics: The origins of a new revolution

Chapter 6: The Theory of Relativity and Quantum Physics

Einstein's theory of relativity and quantum physics are two pillars of modern physics, each revolutionising our understanding of the universe in unique ways. Although they appear to deal with distinct areas of physics, their interactions and divergences have played a crucial role in the evolution of scientific understanding.

Albert Einstein introduced the theory of relativity in two phases: special relativity in 1905 and general relativity in 1915. Special relativity revolutionised our understanding of space and time, asserting that they are relative rather than absolute. It introduced the famous equation $E=mc^2$, which establishes a relationship between energy (E) and mass (m). General relativity transformed our understanding of gravity, presenting it not as a force, but as a curvature of space-time caused by mass.

Quantum physics, developed mainly in the 1920s, deals with the behaviour of particles at the atomic and subatomic scales. This theory revealed a world in which probabilities and indeterminacy are intrinsic.

The theory of relativity and quantum physics approach reality from different perspectives. Relativity applies mainly to the large-scale universe - planets, stars and galaxies - and is exceptionally accurate in describing phenomena involving high speeds and large masses. Quantum physics, on the other hand, is extremely effective in explaining phenomena on the atomic and subatomic scales.

One of the major challenges facing contemporary physics is the quest to unify these two theories, which are based on divergent principles and often appear incompatible, especially in extreme circumstances such as the conditions prevailing within black holes or during the initial instants of the Big Bang.

At the heart of this challenge is the quest for a theory of quantum gravity that would integrate general relativity with quantum mechanics. Several approaches have been proposed, such as *string theory* [1] and *loop quantum gravity* [2], but none has yet been fully accepted or experimentally validated.

Although Einstein was a pioneer of the theory of relativity, he also made a significant contribution to the birth of quantum physics. His work on the photoelectric effect and light quanta laid the foundations of quantum mechanics. However, he was sceptical about certain aspects of quantum mechanics, in particular the idea of indeterminacy and quantum entanglement, which ran counter to his belief in objective reality.

The differences between relativity and quantum physics have also given rise to profound philosophical debates. The principle of complementarity, introduced by Niels Bohr, suggests that relativity and quantum physics, although theoretically incompatible, could be seen as complementary descriptions of reality. This idea has been a subject of intense discussion and debate in the scientific community.

The coexistence of the theory of relativity and quantum physics has greatly influenced modern science. It has led to significant advances in cosmology, astrophysics and the study of extreme phenomena in the universe. The existence of phenomena such as black holes and dark matter can be better understood by integrating concepts from these two theories.

The relationship between the theory of relativity and quantum physics is complex and fascinating. Each offers a unique perspective on the universe, and their interaction continues to stimulate innovative research and philosophical debate. The unification of these two theories remains one of the most exciting challenges in contemporary physics, promising to unveil even more of the mysteries of the universe.

[1] String theory is a field of research in theoretical physics that attempts to describe the fundamental forces and elementary particles of the universe in terms of small vibrating 'strings', rather than dimensionless points.

[2] Loop quantum gravity is a theory that attempts to describe gravity within the framework of quantum mechanics. It is one of the main approaches to a theory of quantum gravity, which aims to unify Einstein's general relativity, which describes gravity on a large scale, with the principles of quantum mechanics that govern subatomic phenomena.

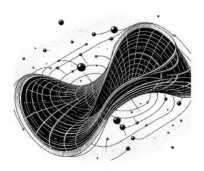

Chapter 7: Applications and Implications of Quantum Physics

By revealing the laws that govern subatomic particles, quantum physics has opened up a vast field of practical applications that have transformed many aspects of our daily lives. From medicine to computing to technology, its implications are vast and far-reaching.

1. Technology and Electronics

Quantum physics has played a crucial role in the development of many technologies that we take for granted today.

Transistors and Semiconductors: The invention of the transistor, the foundation of modern electronic circuits, is a direct result of the understanding of the quantum properties of semiconductor materials. This led to the miniaturisation of electronic components and the advent of the digital age.

Lasers: The laser, a ubiquitous tool in modern life, operates on the principles of quantum mechanics. The stimulated emission of photons, a quantum process, is at the heart of its operation. Lasers have applications in a wide range of fields, from surgery to data communication.

2. Quantum Computing

Quantum computers: Quantum computing is an emerging field that uses the properties of superposition and entanglement of quantum states to perform calculations. Unlike conventional computers, which use bits (0 or 1), quantum computers use qubits, which can be in superposition states (simultaneously 0 and 1). This enables quantum computers to process massive amounts of information at unprecedented speed.

Quantum cryptography: Quantum cryptography uses quantum entanglement and the laws of quantum physics to create virtually unbreakable communication systems. It promises increased security in the transmission of information.

3. Medicine and Imaging

Nuclear Magnetic Resonance (NMR): NMR, used in magnetic resonance imaging (MRI) machines, is based on the principles of quantum physics. It provides detailed images of the body's internal structures without the use of harmful radiation.

Targeted therapies and medicines: Understanding quantum interactions has enabled the development of more targeted drugs and therapies. The design of drugs on a molecular scale and the understanding of chemical interactions in the body are made possible by the principles of quantum chemistry.

4. Materials Science

New materials: Quantum physics has led to the creation of new materials with unique properties, such as superconductors (which conduct electricity without resistance at very low temperatures) and nanomaterials, which have applications ranging from electronics to energy.

5. Energy and Environment

Photovoltaic cells: Solar cells, which convert light into electricity, work thanks to an understanding of the quantum properties of photons and electron-photon interactions.

Studying climate phenomena: Understanding quantum phenomena is also used to model and understand complex phenomena such as climate change.

6. Philosophical and Cultural Impact

Philosophy and metaphysics: Quantum physics has also influenced philosophical and metaphysical thought, calling into question notions of reality, causality and existence.

7. Emerging Technologies

Quantum detection and metrology: The use of quantum systems for measurement and detection is an emerging field. These quantum technologies promise unprecedented levels of accuracy and sensitivity in a variety of applications, from navigation to environmental sensing.

The applications and implications of quantum physics are as vast as they are profound, touching almost every aspect of our modern lives. They extend far beyond technological advances to areas such as medicine, energy, data security and even our understanding of the universe. This branch of science continues to be a source of innovation and discovery, with the potential to further transform our world in the years to come.

History of quantum physics: The origins of a new revolution

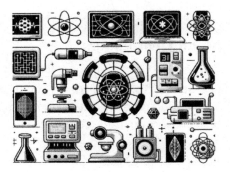

Chapter 8: Challenges and controversies

Since its inception, quantum physics has been the subject of numerous philosophical controversies and theoretical challenges. These discussions often go beyond the strictly scientific to touch on profound philosophical questions about the nature of reality.

1. Indeterminism and Predictability

Probabilistic nature: At the heart of quantum physics is indeterminism, the idea that events at the quantum level are not strictly determined but are instead subject to probabilities. This notion runs counter to the classical idea of an entirely predictable universe.

Einstein-Bohr debate: Einstein's famous phrase, "God doesn't play dice", reflects his unease with the indeterminism of quantum physics. The debate between Einstein and Niels Bohr over determinism and indeterminism was one of the central points of the controversy surrounding quantum mechanics.

2. The Role of the Observer

Measurement and wave-packet reduction: The question of how and why the measurement of a quantum particle 'chooses' one state from many superposed possibilities (wave-packet reduction) is one of the deepest mysteries of quantum physics. This raises questions about the role of the observer and the nature of reality.

Schrödinger's Cat Thought Experiment: This thought experiment was designed to show the apparent absurdity of superposition applied to the macroscopic world, raising questions about the boundary between the quantum and classical worlds.

3. Quantum entanglement and non-locality

Action at a Distance: Quantum entanglement, in which two particles seem instantly to influence each other's state regardless of the distance separating them, defies our intuitive understanding of space and time. It raises questions about the concept of non-locality and the nature of information.

Bell inequalities: Experiments testing Bell inequalities have provided solid evidence of entanglement, challenging notions of hidden local variables and reinforcing the idea of non-locality in quantum physics.

4. Interpretations of Quantum Mechanics

Multiple interpretations: There are several interpretations of quantum mechanics, such as *the Copenhagen interpretation*[3], *hidden variable theories*[4], the *many-worlds interpretation*[5] and *decoherence theory*[6]. Each interpretation attempts to answer the fundamental questions raised by quantum theory in a different way.

Philosophical debates: These interpretations have led to vigorous philosophical debates on questions such as the nature of reality, determinism and the role of consciousness.

5. Limits of Current Understanding

Unification with General Relativity: One of the greatest theoretical challenges is to unify quantum mechanics with Einstein's theory of general relativity, which governs large-scale gravitation. This quest for a theory of everything is at the heart of current efforts in theoretical physics.

Lack of experimental evidence: For some predictions of quantum physics, particularly those involving extremely high energy scales or quantum gravitational effects, direct experimental evidence is still lacking.

6. Impact on other areas

Philosophy and Metaphysics: The questions raised by quantum physics have had a significant impact on philosophy, metaphysics and even fields such as theology and epistemology.

Science and society: The debates surrounding quantum physics also touch on broader questions about the nature of science and its role in society, particularly in terms of public understanding of complex scientific concepts.

Quantum physics continues to pose both theoretical and philosophical challenges. Its paradoxes and mysteries stimulate scientific and philosophical research, and although some of its predictions are already being used in revolutionary technologies, the full potential of this theory remains to be explored. The debates it sparks are enriching not only science, but also our overall understanding of reality.

History of quantum physics: The origins of a new revolution

³ The Copenhagen interpretation is one of the best-known and historically important explanations of quantum mechanics. Formulated in the 1920s mainly by Niels Bohr and Werner Heisenberg, it provides a framework for understanding how quantum observations relate to the world we experience.

4 Hidden variable theories are a category of interpretations of quantum mechanics that suggest that the apparently random and indeterministic behaviour observed in quantum experiments could be the result of unobserved deterministic variables. These variables, often referred to as 'hidden variables', would be unknown or inaccessible to experimentation, but would dictate the behaviour of quantum systems.

5 The many-worlds interpretation (MWI), also known as the multiple universes interpretation, is an interpretation of quantum mechanics proposed by physicist Hugh Everett in 1957. This theory postulates the existence of a very large, even infinite, number of parallel universes that coexist with our own.

6 Quantum decoherence theory is an approach in quantum physics that attempts to explain why we do not see quantum behaviours, such as superposition and entanglement, in the everyday macroscopic world. It proposes a mechanism by which quantum systems lose their quantum properties when they interact with their environment.

Chapter 9: Quantum Physics and Philosophy

Quantum physics, with its non-intuitive concepts and counter-intuitive phenomena, has had a profound impact not only on science, but also on philosophy. It has challenged our traditional conceptions of reality, causality and knowledge, and opened up new avenues in the philosophical understanding of these concepts.

1. Reality and Objectivity

Nature of Reality: Quantum physics suggests that reality on the quantum scale is radically different from our everyday experience. Phenomena such as superposition and quantum entanglement challenge the idea of an objective and stable reality.

Role of the observer: The Copenhagen interpretation of quantum mechanics, which emphasises the role of the observer in determining quantum states, has led to questions about the nature of objectivity and reality itself. It has stimulated debate about whether reality exists independently of observation.

2. Causality and Determinism

Quantum indeterminism: The probabilistic nature of quantum mechanics defies the classical idea of determinism, where the future state of a system can be accurately predicted from its current state. This has profound implications for our understanding of causality.

Non-locality and Action at a Distance: Quantum entanglement and the results of experiments on Bell's inequalities suggest that separate particles can be connected in a way that goes beyond the classical limits of space and time, challenging traditional notions of local causality.

3. Knowledge and understanding

Limits to Knowledge: Quantum physics has revealed fundamental limits to our ability to know certain properties of quantum systems (as illustrated by Heisenberg's uncertainty principle). This has implications for our understanding of what we can know about the world.

Multiple theories and realities: Interpretations such as multiple worlds raise questions about the nature of truth and reality. They suggest that different realities can coexist, challenging our conception of the universe as a single, coherent whole.

4. Metaphysical implications

The question of consciousness: The centrality of the observer in quantum mechanics has led some to speculate on the role of consciousness in determining quantum reality, although this idea remains controversial and largely speculative.

Interconnection and Holism: Quantum physics raises the possibility of a fundamental interconnection of the universe, where the distinct boundaries between entities are less clear, favouring a more holistic vision of reality.

5. Global Philosophical Impact

Science-philosophy dialogue: Quantum physics has strengthened the dialogue between science and philosophy, showing that scientific advances can have profound philosophical implications.

Re-evaluation of the Foundations: It has also forced a re-evaluation of the foundations of the philosophy of science, particularly with regard to the nature of scientific theory, the experimental method and the interpretation of data.

6. Ethics and Responsibility

Quantum technology and society: The advent of technologies based on quantum physics, such as quantum computing and quantum cryptography, raises new ethical questions and societal responsibilities, requiring philosophical reflection on their use and impact.

Quantum physics has opened up new and sometimes bewildering perspectives on fundamental questions of reality, causality and knowledge. Not only has it reshaped our understanding of the physical world, it has also had a lasting impact on philosophical thought, inviting us to rethink our most fundamental conceptions of the universe and our place within it.

History of quantum physics: The origins of a new revolution

Chapter 10: The Future of Quantum Physics

The future of quantum physics promises to continue to reshape our understanding of the universe and inspire technological innovation. Advances in this field promise to be as diverse as they are revolutionary.

1. Unification of Fundamental Theories

Quantum gravity: One of the major challenges remains the unification of quantum gravity with the other fundamental forces, a quest that could lead to a theory of everything. Theories such as loop quantum gravity and string theory are potential candidates, but require experimental proof or further theoretical advances.

Understanding the Big Bang and black holes: A better understanding of quantum gravity could also shed light on the mysteries surrounding the Big Bang and black holes, offering new insights into the origin and evolution of the universe.

2. Quantum Technologies

Quantum computing: The development of fully functional quantum computers could revolutionise fields such as cryptography, optimisation of complex problems and molecular modelling. This would have profound implications for data security, drug discovery and solving complex algorithmic problems.

Quantum communication: Quantum communication, including quantum cryptography, promises virtually unbreakable means of communication. Large-scale implementation of these technologies could radically transform the world's communication infrastructures.

3. Advances in Particle Physics and Cosmology

High Energy Experiments: New experiments, such as those at the Large Hadron Collider (LHC) and other particle accelerators, could reveal unexpected particles and phenomena, offering new insights into fundamental physics.

Exploring Dark Matter and Dark Energy: Quantum physics could play a crucial role in understanding dark matter and dark energy, which make up a large part of the universe but remain mysterious.

4. Interdisciplinary Applications

Medicine and biology: Quantum applications could revolutionise medicine, in particular through the development of new medical imaging techniques and a better understanding of biological processes at the quantum level.

Materials and Nanotechnology: A deeper understanding of quantum phenomena will pave the way for the creation of new materials with novel properties, influencing sectors such as energy, electronics and nanotechnology.

5. Philosophical and ethical challenges

Philosophical implications: As quantum physics continues to challenge our understanding of reality, it will also stimulate dialogue between science and philosophy, particularly with regard to the nature of knowledge, reality and existence.

Ethical considerations: With the development of advanced quantum technologies, ethical issues will emerge, requiring in-depth reflection on their societal impact and responsible use.

6. Education and Society

Dissemination of knowledge: The growing importance of quantum physics will require better integration of its concepts into education, to prepare future generations for the realities of a world increasingly influenced by quantum mechanics.

Public Involvement: Raising public awareness of developments in quantum physics will be essential for a broader understanding and acceptance of quantum technologies and their implications.

The future of quantum physics is rich in promise and challenge. As we continue to explore the infinitely small, the spin-offs from this research should continue to have a significant influence on science, technology and even the structure of society. Quantum physics will remain a dynamic and evolving field, at the forefront of our quest to understand the mysteries of the universe.

History of quantum physics: The origins of a new revolution

44

Epilogue: On the threshold of a new horizon

As we close the pages of this journey through the intricacies of quantum physics, we stand on the threshold of a new horizon, not only scientific but also philosophical, technological and even cultural. In essence, quantum physics is much more than just a branch of physics. It represents a window onto the deepest mysteries of the universe, a mirror reflecting the limits of our understanding and a catalyst for innovation.

From its very beginnings, quantum physics has been a revolution. It has revolutionised the way we perceive the world, from the structure of matter to the foundations of reality itself. It has challenged our most basic intuitions and rewritten the rules governing the universe at the microscopic level. But this revolution is far from over. Each discovery and theoretical advance open up new questions, revealing even more of the complexity and elegance of the quantum universe.

The impact of quantum physics is not confined to the laboratories and blackboards of theorists. Its applications have shaped and will continue to shape our world. From the semiconductors and lasers at the heart of our computers and smartphones, to the medical imaging devices that save lives, quantum physics is everywhere. With the promising advent of quantum computing and quantum cryptography, we are on the cusp of a new technological revolution that could transform our societies in unimaginable ways.

Quantum physics is not just a scientific adventure; it is also a philosophical exploration. It pushes us to question the nature of reality, the role of the observer and even the nature of time and space. The concepts of superposition, entanglement and non-locality challenge our traditional understanding of the universe and suggest a world that is much more interconnected and less deterministic than we had imagined.

What's more, with the development of technologies based on quantum physics, ethical questions are emerging. How will we manage these powerful tools? What measures must we take to ensure security, confidentiality and fairness in a world influenced by quantum computing and quantum surveillance?

The future of quantum physics will largely depend on our ability to educate future generations. It is essential to integrate these concepts into our education systems, not only to train tomorrow's physicists, but also to enlighten citizens about a world that is increasingly influenced by quantum science. This will require a concerted effort to make quantum physics accessible and understandable, moving beyond jargon and complex formulae to appeal to the imagination and curiosity of all.

In conclusion, quantum physics is not a final destination, but an ongoing journey, a path strewn with questions as fascinating as they are complex. It invites us to look at the world not as something fixed and determined, but as a web of possibilities, where reality is a dynamic puzzle in constant evolution. As we pursue this eternal quest for understanding, quantum physics will remain a source of wonder, a challenge to our intellect, and a guide to the unexplored possibilities of the universe.

History of quantum physics: The origins of a new revolution

References and additional resources

For those who wish to deepen their understanding of quantum physics and further explore its concepts, applications and philosophical implications, the following list offers a selection of bibliographical references and additional resources. These works have been chosen for their relevance, accessibility and ability to shed light on different aspects of quantum physics.

1. Books and articles

Quantum mechanics - The theoretical minimum by Leonard Susskind and Art Friedman - An accessible guide to understanding the foundations of quantum mechanics.

Helgoland: The Strange and Beautiful Story of Quantum Physics by Carlo Rovelli - Understanding quantum physics.

Quantum Computing: An Applied Approach by Jack D. Hidary - his book integrates the foundations of quantum computing with a hands-on coding approach to this emerging field.

Something Deeply Hidden: Quantum Worlds and the Emergence of Spacetime by Sean Carroll - An authoritative and beautifully written account of the quest to understand quantum theory and the origin of space and time

The Little Book of String Theory by Steven Gubser - A book that explores string theory and its potential role in unifying the forces of the universe.

Articles from specialist journals such as Physical Review Letters, Nature Physics, and Science, offering research and discussion at the cutting edge of quantum physics.

2. Online resources

arXiv.org - An open-access archive containing pre-publications of research work in the fields of physics, mathematics and computer science.

The Quantum Atlas - A website dedicated to exploring the concepts of quantum physics through interactive visualisations and detailed explanations.

TED talks on quantum physics - Presentations by leading physicists and thinkers in the field, available on TED.com.

Online courses such as those offered on Coursera and edX, offering university-level courses in quantum physics taught by renowned institutions.

Quantum Mechanics - MIT OpenCourseWare - A comprehensive quantum physics course from MIT, including video lectures, lecture notes, and exercise problems.

3. Documentaries and Educational Videos

The Fabric of the Cosmos by NOVA - A documentary series exploring the nature of space, time and quantum reality.

YouTube channels such as PBS Space Time, Veritasium and MinutePhysics, offering clear and engaging explanations of

History of quantum physics: The origins of a new revolution

various topics in quantum physics.

In addition to these resources, it is also advisable to follow the latest publications in scientific journals and attend conferences or seminars on quantum physics, as the field is constantly evolving with new discoveries and emerging theories.

ABOUT THE AUTHOR

Coming from a background steeped in science, I am perpetually fascinated by the captivating mysteries of the universe. My insatiable curiosity, which goes far beyond my formal education, constantly drives me to seek answers and deepen my understanding.

I am truly captivated by the richness and diversity of the subjects that make up the complex fabric of our existence. My aspiration to unravel the mysteries of the universe is both ambitious and exhilarating. However, I am fully aware that the limited span of our human lives restricts our ability to decipher all its enigmas and to answer all the questions we face.

This reflection on the breadth of as yet unexplored knowledge influences my approach on a daily basis. Every day is a new opportunity to broaden my knowledge, question my beliefs and marvel at the hidden splendours of our universe.

In this spirit, this book represents my modest but fervent contribution to a field of study that is both vast and captivating. Its aim is not only to shed light on certain little-known aspects, but also to share my thoughts and analyses on the dynamics of the world around us. I hope that these pages will encourage further exploration of as yet unsolved mysteries and spark the intellectual curiosity of my readers, inviting them to join in this ongoing quest to understand our dynamic and ever-changing reality.

Printed in Great Britain
by Amazon